MW01657083

Sciencepalooza

A Collection of Science Poetry for Primary and Intermediate Students

Franny Vergo

authorHOUSE®

AuthorHouse™
1663 Liberty Drive
Bloomington, IN 47403
www.authorhouse.com
Phone: 1-800-839-8640

© 2012 Franny Vergo. All rights reserved.

No part of this book may be reproduced, stored in a retrieval system, or transmitted by any means without the written permission of the author.

Published by AuthorHouse 7/10/2012

ISBN: 978-1-4772-1598-2 (sc)
ISBN: 978-1-4772-1597-5 (e)

Library of Congress Control Number: 2012910052

Any people depicted in stock imagery provided by Thinkstock are models, and such images are being used for illustrative purposes only.
Certain stock imagery © Thinkstock.

This book is printed on acid-free paper.

Because of the dynamic nature of the Internet, any web addresses or links contained in this book may have changed since publication and may no longer be valid. The views expressed in this work are solely those of the author and do not necessarily reflect the views of the publisher, and the publisher hereby disclaims any responsibility for them.

Sciencepalooza is dedicated to all the inquisitive science students I had the privilege of teaching for thirty-six years!

Contents

I. INTRODUCTION

WHAT IS THE METHOD OF THE SCIENTIFIC METHOD?

What is the method of *the scientific method?* What do scientists do?
It's a series of steps that scientists follow for an answer they pursue.
It starts with a QUESTION or PURPOSE of the investigation.
Then the scientist does RESEARCH and makes careful OBSERVATIONS.
Next, they form a HYPOTHESIS for the answer to their question,
Based on their research and observations—it's a guess or prediction.
An important step is to TEST this guess with an EXPERIMENT;
Scientists really love this part and do it with great merriment.
Next, they ANALYZE all their DATA and DRAW A CONCLUSION,
Then COMMUNICATE their results clearly with no confusion.

INVESTIGATE AN INVESTIGATION

Let's investigate a science investigation,
And make a very careful examination.
It's the scientific method's *experiment* part,
A very vital step because it is the heart.

First, DESIGN the EXPERIMENT SETUP
And GATHER MATERIALS needed, like a cup.
Next, decide what the VARIABLE will be.
This one main difference is the answer's key.

Always RECORD the DATA that you observe
To get the right RESULT and not be thrown a curve.
Last, remember science SAFETY during your investigation.
Take all precautions over the experiment's duration.

LAB TOOLS RULE

The tools we use in a science lab totally rule
Because they are helpful, awesome, and cool.
The TEST TUBE, the BEAKER, and the FLASK
Hold and measure liquids till you complete your task.
A GRADUATED CYLINDER measures different objects' volumes.
SPECIAL SPOONS measure CHEMICALS a scientist never consumes.

Use a BALANCE SCALE to measure mass, a WEIGHT SCALE for weight.
Heat substances on a one-flame BUNSEN BURNER or HOT PLATE.
Use a PETRI DISH to grow bacteria at its own pace,
And LITMUS PAPER to identify an acid or base.
Measure temperature with a THERMOMETER, a useful tool.
All these awesome science lab tools totally rule.

FIRST AID KIT
FIRE
NGUISHER

BE SAFE WITH SCIENCE SAFETY

Always be safe with science safety, kids,
When experimenting at home or at school.
Use special materials and safety apparel.
Always follow every lab safety rule.

When needed, wear a pair of safety GOGGLES,
A strong, protective APRON, and safety GLOVES.
Store a FIRE EXTINGUISHER and a good FIRST-AID KIT.
Be careful and safe doing what a scientist loves.

Follow these important lab safety rules:
STAY FOCUSED and HAVE AN ADULT WITH YOU.
Also, NEVER TASTE OR DRINK ANYTHING
And WASH YOUR HANDS THOROUGHLY when you're through.

ARE THE MAD SCIENTISTS REALLY MAD?

Are the mad scientists really mad?
Well, mad about science, and that's not bad.
They're CURIOUS and OBSERVANT every day,
Also, totally HONEST in every way.
They're PERSISTENT because they never give up
And OBSESSIVE in getting their answers—yup!
So, are the mad scientists really mad?
Okay, maybe they are, but just a tad.

I am a BOTANIST!
I am a PALEONTOLOGIST!
I am a ZOOLOGIST!
I am a GEOLOGIST!

SCIENTISTS GALORE

Many types of scientists exist,
That are always honest and persist.
They perform science studies—that's the gist.
Here's a sample alphabetical list:

ARCHEOLOGISTS study cultures of the human race.
ASTRONOMERS probe everything that is in space.
BIOLOGISTS observe living organisms and their growth pace.
BOTANISTS are experts in plant stages, whatever the case.
CHEMISTS analyze compositions of all types of matter.
CYTOLOGISTS examine cell structures—some thinner, some fatter.
GEOLOGISTS study the earth's minerals and rocks, bumpy to flatter.
METEOROLOGISTS report the weather, like drizzle's pitter-patter.
PALEONTOLOGISTS dig for fossils and study dinosaur life.
PHYSICISTS see how matter and energy relate, sometimes with strife.
SEISMOLOGISTS measure earth's movements—earthquakes split the
ground like a knife.
ZOOLOGISTS group animals and learn about their habitat life.

II. SPACE SCIENCE

THE COSMIC COSMOS

The cosmic cosmos is cosmic indeed.
Expanding outward, growing like a weed.
The totality of EVERYTHING THAT EXISTS.
All MATTER AND ENERGY that persists.
Billions of GALAXIES, INTERGALACTIC SPACE.
The humongous universe rotates with grace.

GALACTIC GALAXY SHAPES

Galactic galaxies are classified by shape,
Gravity in and around them so they don't escape.
There are three main galactic shapes in space.
Each made of stars, rotating at their own pace.

First is an ELLIPTICAL galaxy with an oval form.
Some are nearly spherical in shape, but it's not the norm.
It is the largest galaxy type in the universe
With mostly older, evolved stars and few star births.

Next is the SPIRAL galaxy, like our own Milky Way.
It's pinwheel-shaped with graceful arms where new stars form and stay.
A lot of dust between the arms, but not one single star.
Some spirals have a center band of stars shaped like a bar.

An IRREGULAR kind has no shape at all.
This type has no special form and is usually small.
It's formed by the gravity pull of galaxies nearby.
Irregulars are small, with gravity in short supply.

OH, MILKY WAY GALAXY

Oh, Milky Way galaxy seen from afar,
Are you named after the famous candy bar,
Or is the sweet chocolate bar named after you?
This is a fact many people wish they knew.

Sometimes there's a large, milky white band of light
Made from distant stars arching across the sky at night.
More than two hundred billion stars with a fiery glow,
That's how you got your name, and now people know.

But was the candy bar named after you?
Some people want to know this fact too.
The answer is quite interesting. You see,
It was named for a malt in 1923.

Jupiter
Asteroid Belt
Mars

AMAZING ASTEROIDS

Amazing asteroids are large objects made of METAL AND ROCK,
That never formed into planets, but *rocky* planets they mimic or mock.
Most are between Mars and Jupiter, orbiting the Sun.
In this huge *asteroid belt,* thousands are on the run.

THE THREE ASTRONOMICAL Ms

The three astronomical Ms have the same composition.
The difference between them is their location and position.

The first astronomical *M* is a METEOROID:
It's a piece that broke off a larger asteroid.

The second astronomical *M* is a METEOR you see from afar:
A burning meteoroid falling through Earth's atmosphere—a *shooting star.*

The last astronomical *M* is a METEORITE:
It's a meteor that's made it to Earth in plain sight.

Three astronomical Ms that are all made the same;
Each in a different location with a different name.

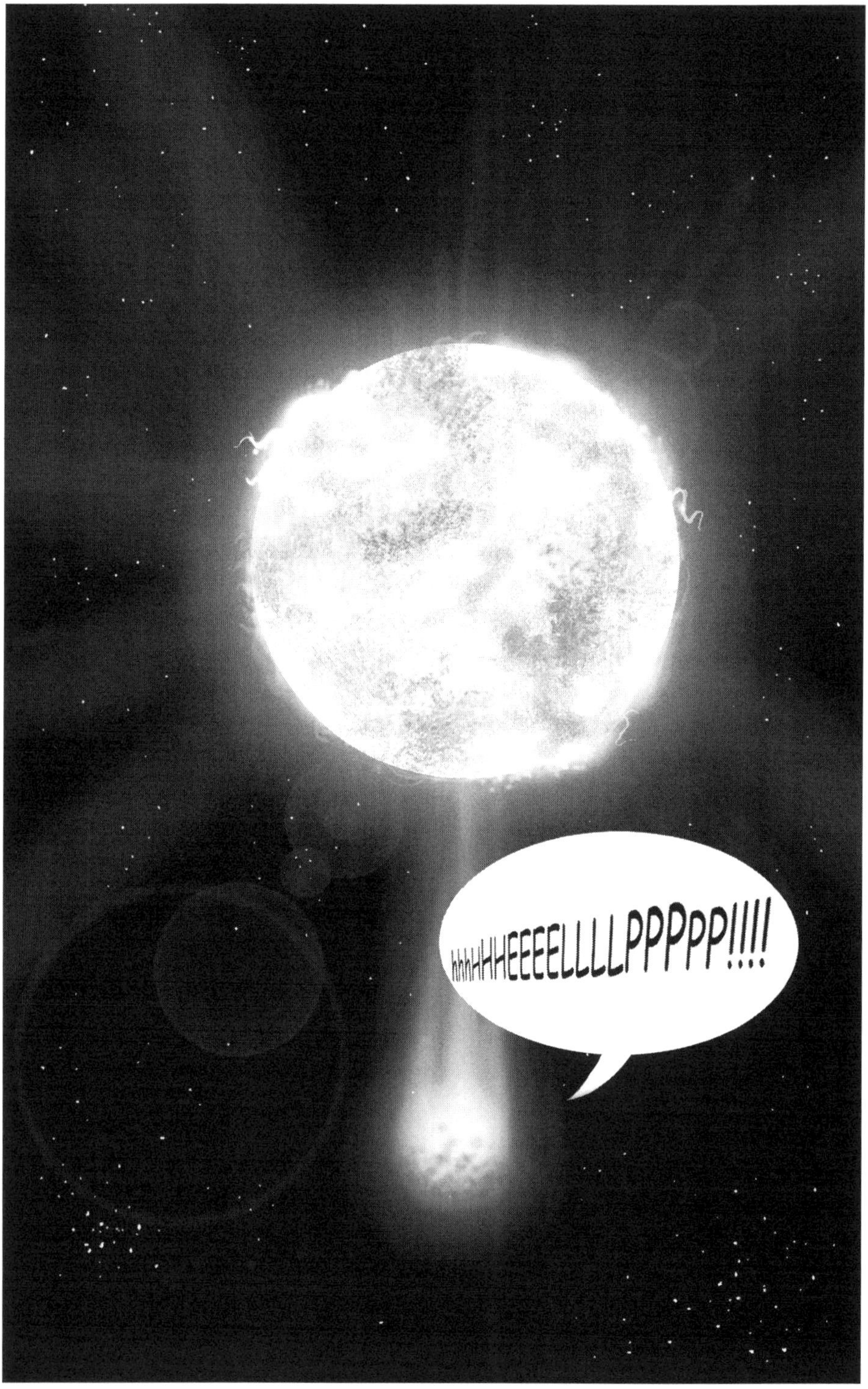
hhhHHEEEELLLLPPPPP!!!!!

COMET, COMET

Comet, comet, flying in your orbit every day
Around the Sun, your long tail streaming away.
Your ICE HEAD and GAS TAIL are visible when you're near the Sun.
But don't get too close, or you'll melt, and then you'll be done.

HERE COME THE EIGHT PLANETS

Here come the eight planets
Revolving around our big Sun.
Here come the eight planets
Coming around one by one.

Learning about space is so great,
Especially the planets that are first-rate.
From Mercury to Neptune, you will see
That the planets are as different as they can be.

The inner planets are the closest to the Sun.
They're made of rock and weigh lots more than a ton.
Mercury and Venus are the very first two,
Followed by Earth, then planet Mars that we pursue.

The outer planets are the next in line.
Big Jupiter and Saturn have rings so fine.
Then Uranus on its side and the planet Neptune.
These four are made of gases and are frozen through June.

There go the eight planets
Revolving around our big Sun.
There go the eight planets
Orbiting around one by one.

I'm Pluto, and now I'm just a Dwarf Planet!

POOR LITTLE PLUTO

Everyone feels sad and sorry for poor little Pluto
'Cause he was demoted to a DWARF PLANET some years ago.
Pluto is not an official planet anymore
'Cause he shares his orbit; alone, he does not soar.
Little Pluto no longer is one of the nine
'Cause scientists kicked him out. He did not resign.
NASA's space probe *New Horizons* was sent to study him;
It will capture images that for now are just prelim.

I LOVE THE MOON

I love the Moon. I love the Moon. Oh, how I love the Moon!
If I had the chance to go there, it wouldn't be too soon.
The beautiful, bright moonlight on a summer eve's stroll.
The *man in the moon* looking down, smiling with his droll.

The awesome ocean tides washing over the sandy beach.
Each day, low tides and high—the latter make a farther reach.
The Moon's gravitational pull helps to create these tides
That are stronger when there's a full moon that overrides.

When there's a full moon, mammal behavior might change.
When the Moon is full, animals can act a little strange.
The full Moon also creates the perfect setting for love.
People are *moonstruck* from the light shining from above.

Life would be different without our friend, the Moon,
Which you can sometimes see on a clear afternoon.
And it's fun to watch the phases as they wane and wax.
I love the Moon. Oh, I love the Moon to the very max.

SPACE JUNK

There is a lot of space junk orbiting our Earth every day.
Many thousands of space objects are in the way.
This junk includes rocket bodies and mission-related debris,
And some nonfunctional spacecraft, fuel, paint flakes, batteries.
A good, practical solution needs to be found on the double
'Cause all this junk cluttering space might cause some trouble,
Like jeopardizing space travel and future space missions,
With the risk of operational spacecraft collisions.
It's bad enough that planet Earth has way too much junk and debris.
We need to keep our planet and the space above it clutter-free.

OFF TO THE SPACE STATION!

I WISH I WERE AN ASTRONAUT

I wish I were an astronaut in a flight bound for space,
In an aerodynamic spacecraft flying up with grace.
Once out of Earth's atmosphere, with no gravity,
I'd float around upside down. How fun it would be!

I'd see satellites and the huge Hubble telescope,
And I wouldn't crash into the floating junk, I hope.
I'd dock my spacecraft and go aboard the space station,
And conduct experiments to learn new information.

When back on Earth, I'd share all that I learned and saw
And explain that while in space, I was in total awe.
I'd teach people the importance of space travel
'Cause space is a mystery we need to unravel.

THE LIFE CYCLE OF STARS

A star is a burning ball of hydrogen and helium gas,
Whose life span can be many billions of years.
All stars are classified by their LIFE-CYCLE STAGE,
And their size, mass, and temperature color that appears.

Stars have a long life cycle from their birth to their death.
It starts in a NEBULA cloud of thick dust and gas,
In a region of dense matter called a PROTOSTAR.
Then nuclear reactions make a MAIN-SEQUENCE STAR's mass.

Most stars in the universe are in their main-sequence stage.
Nuclear fusion starts the fire, sending out light and heat too.
Burning starts when the hydrogen turns to helium.
When the hydrogen runs out, nuclear fusion is through.

Then the star begins to die in a couple of new stages.
SMALLER STARS expand out to cooler GIANTS THAT ARE RED.
As they expand, they form NEW NEBULA for more star births.
Then shrink to a WHITE DWARF, and lastly, a BLACK DWARF that is dead.

When LARGER STARS die, they become RED SUPERGIANTS.
Then there's a SUPERNOVA or very large explosion.
The leftover matter is now a dense NEUTRON STAR,
Or a super dense BLACK HOLE that has incredible suction.

MEET OUR SUN

Meet our Sun, the CLOSEST STAR TO EARTH,
The CENTER OF OUR WHOLE SOLAR SYSTEM.
The only star we can see in the day,
The star that all earthlings daily welcome.

It's a glowing BALL OF BURNING GAS
Made up mostly of HYDROGEN,
And a small amount of HELIUM GAS
That our Sun gets from nuclear fusion.

Our own Sun is a MEDIUM-SIZED STAR,
And has a MEDIUM TEMPERATURE too.
That is why its color is bright YELLOW,
Not deep red, orange, white, or blue.

The Sun's center is the superhot CORE.
Next is the Sun's surface or PHOTOSHPERE.
The last part is the CORONA region,
The Sun's hot outer atmosphere.

The Sun SUPPORTS ALL LIFE ON PLANET EARTH:
It gives us the LIGHT AND HEAT that we need.
Without it, green plants could not make their food,
So there would be no life on Earth indeed.

But please do not worry or be concerned.
Our sun will live five billion years afar.
Then into a cool red giant it will swell,
And shrink to a small white, then black dwarf star.

BACK OFF, BIG BLACK HOLE

Back off, big black hole. Please do not get near me
Because I'll get sucked up by your strong gravity.
And once I get pulled in, I will never get out,
Even if I jump high and kick or scream and shout.
In your intense, compact mass, I will surely be squished
Together with trapped light, and that's surely not my wish.
My wish is that you always stay billions of light-years away,
Where your intense gravitational field will always stay.
So, back off, big black hole, and don't even think about it!
Just stay far away from me 'cause I don't like you one bit!

STARRY STARS

Although stars look white from Earth, they all have a different hue.
The coolest stars are deep red, and the hottest stars are blue.
After the hottest blue, there's white, yellow, orange, and red,
And of course, there are no colors when the stars are dead.

The Sun is the closest star by far to planet Earth.
The star *Proxima Centauri* is next, for what it's worth.
The brightest star from Earth is *Sirius,* besides our Sun;
Though fifth in distance from our Earth, it's a hot, white one.

A star's brightness is the sum of energy sent to Earth;
Its size, mass, distance, temperature, and age from birth.
And though we say, "Stars twinkle," they really do not.
Their light bends above Earth 'cause of the atmosphere we've got.

All stars in the universe live billions of years,
Moving through their life-cycle stages in their hot spheres.
Giant stars have a shorter life span than the smaller stars
Because they burn hotter and faster than a star like ours.

STARRY, STARRY NIGHT

It's so awesome to gaze at the sky on a dark, starry, starry night,
Where there are more than one thousand beautiful stars in clear sight.
It's fun to connect the stars and find pictures in the night sky;
These CONSTELLATIONS are intriguing to watch, as each month they pass by.

The ancient people who created constellations made up stories too,
Like Greek myths with gods and goddesses, who could cause a hullabaloo.
Ancient farmers used the constellations to learn the time of year,
To know when it was time to plant or reap and get in gear.

These ancient people found forty-eight constellations in their night sky,
And today, there are eighty-eight known constellations to view up high.
We use these star formations to help us locate many of the stars,
And to help us understand mythology, which can be quite bizarre.

There are famous CONSTELLATION FAMILIES that you can identify.
They are related in a myth and are close together in the sky,
Like the *Perseus Family*, which includes *Pegasus*, the flying horse,
Or the *Orion Hunting Family* with *Big and Little Dog,* of course.

Start finding some star formations in the sky on a dark, clear night.
Try to go to a good spot without tall buildings or a lot of light,
Have fun looking for constellations that each tell an ancient story.
Your search will bring you joy and knowledge that's hunky-dory.

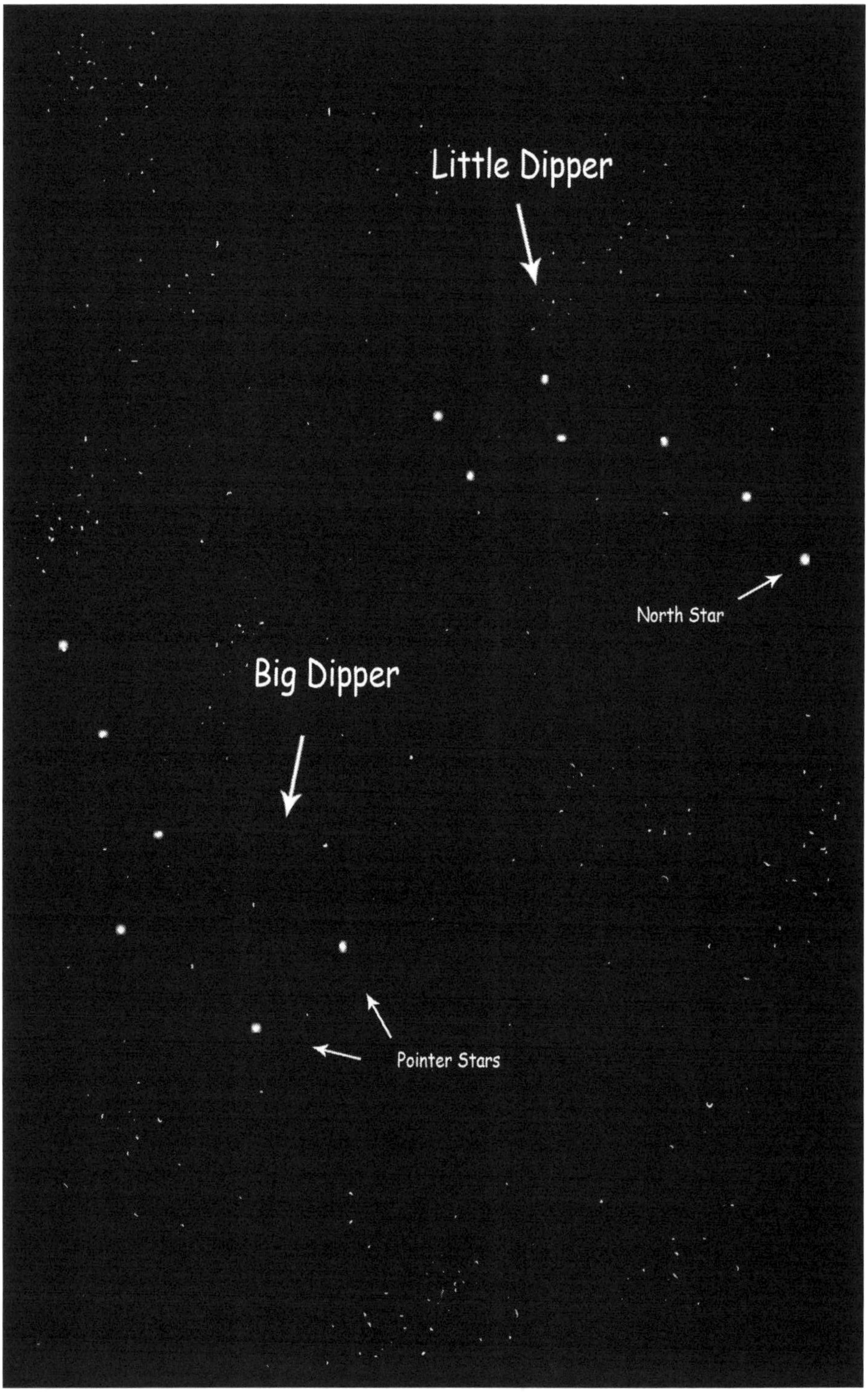
Little Dipper
North Star
Big Dipper
Pointer Stars

DIP INTO THE DIPPERS

Let's dip into the two star dippers and see what facts we can learn.
They both circle around the north celestial pole as they turn.
The two most well-known star formations and the easiest to locate:
The smaller dipper upside down over the bigger as they rotate.

The LITTLE DIPPER way up north is an important CONSTELLATION.
The star at the end of its handle has been used for navigation.
For thousands of years, this NORTH STAR has helped people find their way.
It is in the exact same position every night and day.

Seven stars make up the Little Dipper, which is part of LITTLE BEAR.
You'll see this *asterism* if you look carefully and stare.
The Little Dipper is one third the size of the spread-out Big Dipper,
Always looking like its little cup is pouring into the bigger.

The BIG DIPPER has the same number of stars, but they're farther apart.
And there's a "companion star" by the handle, next to the last star part.
The Big Dipper's also an *asterism* 'cause it's part of BIG BEAR.
The Big Bear is always close to the Little Bear in their den or lair.

The Big Dipper is very large and easy to find in the night sky,
And it helps you find the Little Dipper that is above it nearby.
Follow the two POINTER STARS at the cup's far end and look straight up.
Then you'll find the North Star at the handle's end, to the right of its cup.

III. LIFE SCIENCE

I'LL MAKE MY OWN FOOD, THANK YOU VERY MUCH

I'll make my own food, thank you very much,
Because I am a PRODUCER GREEN PLANT.
I am not a consumer or an animal,
So I don't need to find food, and besides, I can't.

I MAKE MY OWN SUGAR called GLUCOSE, which is my food,
USING CHLOROPHYLL, CO2, WATER, and SUNLIGHT.
My glucose helps me to be very strong and healthy.
PHOTOSYNTHESIS happens in the day, never at night.

I'm the START OF EVERY FOOD CHAIN AND WEB, large or small,
Because I'm the only life form that makes my own food.
Without me, there'd be no food or life on Earth at all,
But I'll save you, so there's no need to worry or brood.

CHAINS AND WEBS OF FOOD

All ecosystems have many chains and webs of food.
The chains and webs BALANCE EVERY ECOSYSTEM'S LIFE.
They AVOID OVERPOPULATION AND EXTINCTION.
It is NATURE'S WAY, though it sometimes causes strife.

A FOOD CHAIN CONNECTS ONE LIFE FORM TO ANOTHER;
EACH ONE PASSES ENERGY TO ONLY ONE LINK.
They ALWAYS BEGIN WITH A PRODUCER, A GREEN PLANT.
If ONE LINK DIES, SO DOES THE CHAIN, but first, it will shrink.

When FOOD CHAINS CONNECT together, FOOD WEBS ARE FORMED;
EACH ONE PASSES ENERGY TO MORE THAN ONE LIFE TYPE.
If ONE PART OF A WEB DIES OUT, the WEB STILL GOES ON
Because the LINK EATS SOMETHING ELSE and just makes a swipe.

We need
YOU!
We
NEED
you!

WE NEED EACH OTHER

We all need green plants, and green plants need us;
This is very important to discuss.
ANIMALS AND PLANTS are PARTNERS IN LIVING
Because of THE GAS EACH OF THEM IS GIVING.

ANIMALS BREATHE IN OXYGEN and EXHALE CO2,
The most critical part of survival that we do.
GREEN PLANTS TAKE CO2 AND GIVE OXYGEN, and we can conclude,
That THIS HAPPENS DURING PHOTOSYNTHESIS WHEN THEY MAKE
THEIR FOOD.

Without this crucial process, green plants would die,
And all other life forms would then go awry
Because we need each other to stay alive,
And also to be healthy, strong, and to thrive.

WHAT DOES THAT ANIMAL EAT?

"What does that animal eat?" you ask.
It depends on the type and its eating task.
Is it CARNIVORE, HERBIVORE, or OMNIVORE?
If you know the type, you learn a lot more.

CARNIVORES CHASE ANIMALS TO EAT.
They're PREDATORS, with the creatures they meet.
The ANIMALS THEY CATCH AND EAT are called their PREY.
Carnivores EAT MEAT, so plants are safe all night and day.

Animals that EAT PLANTS ONLY are HERBIVORES.
They don't chase prey, so their meal is not a chore.
Herbivores go up to plants and munch, munch, munch,
But THEY CAN BE A MEAL FOR PREDATORS! Yikes! Crunch, crunch, crunch.

OMNIVORES EAT ANIMALS AND PLANTS, so they can rejoice.
Whenever they are hungry, they have a food choice.
If they choose to hunt an animal, they can chase their prey,
Or just chomp on some plants and have their meal the easy way.

CONSUMER ORGANISMS is one way to group these three
Because THEY CAN'T MAKE THEIR OWN FOOD like producers, you see.
They must LOOK FOR OTHER FOOD to stay alive.
EATING PLANTS OR MEAT helps these animals survive.

IS THAT ANIMAL A VERTEBRATE OR AN INVERTEBRATE?

Is that animal a vertebrate or an invertebrate?
Let's look at some information and get these two types straight.
We need to know the difference between these two,
So here are some interesting facts to chew:

A VERTEBRATE is an animal that has a BACKBONE,
A rodent, a reptile, or bird, and they're not alone.
A mammal, a primate, a fish, an amphibian too.
A cute marsupial with a big pouch, like a kangaroo.

An animal that is an INVERTEBRATE has NO BACKBONE at all,
A worm, a spider, a scorpion, and insects that fly or crawl.
A lobster, a crab, or a squid in the big ocean blue.
A snail, a clam, or a jellyfish in the water too.

Although the percent of invertebrates is more than ninety-eight,
The size and mobility of vertebrates helps them dominate.
In every ecosystem, the two types live side by side.
They are both part of the great food web in which they reside.

I LOVE YOU, PRAIRIE GRASS

(This poem can be sung to the "Spiderman" tune.)

Prairie grass, prairie grass,
You are all really a gas.
Indian grass and wild rye,
I always hate to say good-bye.
Prairie grass, you are so lovely and cool!

Prairie grass, prairie grass,
Standing in a large green mass.
You are so green, you are so tall,
Except in winter and the fall.
Prairie grass, you are so lovely and green!

Prairie grass, prairie grass,
All other grass you surpass.
Big blue stem up to eight feet tall,
And three feet are the ones that are small.
Prairie grass, you are so lovely and high!

Prairie grass so tall and green,
The most beautiful grass I've ever seen.
When I observe you in the prairie,
It amazes me how much you vary.
I love you! I love you, prairie grass!

SAVE THE PRAIRIE

(This poem can be sung to the "Take Me Out to the Ball Game" tune.)

Take me out to the prairie.
Take me to the tall grass.
Show me the snakes and amphibians too,
The mammals, the birds, and the wild forbs of blue.
Oh, we must try to save the prairie.
If we cut them all down it's a shame.
We need to respect and save the prairie
And each prairie food web and chain.

Yummy!!

THE PRAIRIE BEE

The prairie bee, the prairie bee.
What does it see? What does it see?
Beautiful colors of each prairie forb.
Each has a fragrance it will absorb.
Each full of nectar that's oh so sweet.
For the prairie bee, each forb's a treat.
Though the prairie bee doesn't know it's great,
It is, for each day it will pollinate.
Important pollination to make a new seed,
To make more green plants that we all need.
Keep doing a great job on the flower parts.
We thank you, prairie bee, from the bottom of our hearts!

THE MONARCH BUTTERFLY OF THE PRAIRIE

The monarch flies through the prairie, oh so free,
Looking for sweet nectar with each bird and bee.
It's first a caterpillar and next a cocoon.
Then it turns to a beautiful butterfly soon.
Its job is to pollinate the wildflowers.
It likes to pollinate for many hours.
The monarch butterfly of the prairie,
It flies so lightly like a fairy.
The monarch butterfly of the prairie,
It is so peaceful and so merry.

THE GREEN PRAIRIES OF THE MIDWEST

The settlers moved west to have a good farm life,
But there were problems in the prairie that caused some strife.
The mosquitos and blizzards made the settlers' lives rough,
And all of the plant roots were deep, fibrous, and tough.
Then John Deere came along with a great farm invention.
His steel plow cut those roots smoothly without much tension.
McCormack's reaper made harvesting easier, no doubt,
And J. F. Glidden's barbed wire kept pesky animals out.
The settlers stuck it out and worked hard every day.
These inventions helped a lot, so they decided to stay.

PLEASE KEEP OUR EARTH CLEAN!

IV. ENVIRONMENTAL SCIENCE

ALL THE ANIMALS LIVE HERE TOO

All the animals live here too.
It's not only humans who do.
They need us to keep the Earth clean,
Also beautiful, lush, and green.
They live in each community,
And they're part of our Earth family.
So, let's respect the Earth in all we do
Because all the animals live here too.

I LOVE TREES

I love trees. Oh, I love trees,
Even though some make me sneeze.
In their shade, I love to sit,
Lie back and relax a bit.

Gaze in awe at their splendor.
Seedlings, too, that are slender.
Pick a fruit off with a reach:
An apple, pear, or a peach.

But mostly, I just have fun,
Like when I climb up on one,
Or in my tree house I play
With all my friends every day.

PLANT A PLANT

Save our earth and plant a plant.
Please don't say that you can't.
They're good for the earth, you know,
In lots of ways when they grow.

Plant flowers, bushes, or trees.
Get down and dig on your knees.
Put some seeds into the ground,
Or a seedling that you've found.

All green plants help planet Earth.
In many ways, they have worth.
If there were no plants of green,
Life on Earth would not be seen.

Plants take CO2 from air,
And put oxygen back there.
They provide shelter and food
To help all life get renewed.

Their roots hold Earth's soil in place,
As they stand upright with grace.
They add beauty to each scape,
With each plant's color and shape.

So go out and plant a plant.
Don't complain and say you can't.
There's not much that plants demand:
Just water, light, and good land.

PLEASE STOP FILLING OUR LANDFILLS

Please stop filling our landfills, won't you?
They are filling up way too fast,
With so much different garbage
That together we've amassed.

Many millions of tons of garbage
Are disposed of every year,
Taking up space in landfills,
Causing great concern and fear.

Poorly run landfills release toxins
From rotting and hazardous debris.
Leachate in groundwater,
And methane gas in air, you see.

Too much trash goes to our landfills.
They are so badly abused.
The saddest part of it all
Is that much can still be used.

So, please reduce, reuse, and recycle,
And compost biodegradables in a flash.
Bring hazardous waste to a collection center,
And throw the rest into the trash.

POLLUTION SOLUTION

Pollution pollutes LAND, WATER, and AIR.
It makes them all DIRTY, and that's not fair.
We live with it, drink it, and breathe it.
We see it, feel it, taste it, and smell it,
And we can even hear it too.
It's called NOISE pollution. It's not new.

WE USE TOO MUCH OF EVERYTHING,
Like products with too much PACKAGING
And GROCERY ITEMS that we buy in mass.
Sometimes we need to take a pass.
The tons of PAPER we use every day,
And the TREES that give their lives in this way.

We use up too much of our ENERGY,
NATURAL GAS and ELECTRICITY.
Too much PLASTIC and GASOLINE
Uses up OIL and makes our air unclean.
Every day, more GARBAGE is produced.
Plus, our WATER use should be reduced.

Our pollution problems will never end
If we don't change our ways and bend.
TOO MUCH OF EVERYTHING CAUSES POLLUTION;
CONSUMING LESS IS THE BEST SOLUTION.
We must stop and think before we act,
And make GREEN DECISIONS that we'll enact.

THE THREE Rs

The first *R* is REDUCE, or cut back on our trash.
Buy products with less packaging and save some cash.
The most important *R* for you to use,
So do it often and don't refuse.

The second *R* is REUSE your things now and then.
Instead of throwing them out, just use them again.
Use these things yourself or have a big garage sale
'Cause your trash could be someone's treasure to avail.

RECYCLE aluminum, paper, and most plastic
'Cause giving these things a second life is fantastic.
This third *R* is easy for anyone to do.
Just toss them in your recycle can when you're through.

Doing these three Rs saves money and energy,
And natural resources to a large degree.
It's a great way to be *green* and to help out.
It will help save our planet. There is no doubt.

THE OTHER THREE Rs

Reduce, reuse, recycle to be *green*.
And help keep our earth healthy and clean.
But there are three other Rs that you need
To keep our planet beautiful indeed.

The first is to RETHINK, or stop and think twice.
Before tossing trash out, take this advice:
Use it again if you possibly can,
Or recycle it back to how it began.

The second is to REJECT, or just say, "No!"
To anything that harms our earth, say, "Whoa!"
If it's not healthy or it isn't *green*,
Do not take part, but rather, intervene.

The third is to RECLAIM Earth as your own.
The special planet that you call your home.
If you reclaim it, you'll want to be *green*
Because you'll want your home to be pristine.

HAPPY EARTH DAY
HAVE A SMALL CARBON FOOTPRINT!
GO GREEN!

HAPPY EARTH DAY

(This poem can be sung to the "Happy Birthday" tune.)

Happy Earth Day to you.
Happy Earth Day to you.
Mother Earth, how I love you.
Happy Earth Day to you.

I'll FOLLOW ALL THE Rs.
I'll follow all the Rs.
I'll follow the Rs 'cause I love you.
I'll follow all the Rs.

I'll COMPOST MY GOOD TRASH.
I'll compost my good trash.
I'll compost 'cause I love you.
I'll compost my good trash.

I'll CONSERVE RESOURCES.
I'll conserve resources.
I'll conserve 'cause I love you.
I'll conserve resources.

I'll SAVE ENERGY.
I'll save energy.
I'll save energy 'cause I love you.
I'll save energy.

I'll NEVER LITTER.
I'll never litter.
I'll never littler 'cause I love you.
I'll never litter.

I'll TRADE MY TOYS AND BOOKS.
I'll trade my toys and books.
I'll trade 'cause I love you.
I'll trade my toys and books.

I'll VOLUNTEER TO HELP.
I'll volunteer to help.
I'll volunteer 'cause I love you.
I'll volunteer to help.

I'll PLANT A TREE IN SPRING.
I'll plant a tree in spring.
I'll plant a tree 'cause I love you.
I'll plant a tree in spring.

I'll HAVE A SMALL FOOTPRINT,
A SMALL CARBON FOOTPRINT.
I'll keep it small 'cause I love you.
I'll have a small footprint.

I'll READ BOOKS ABOUT YOU.
I'll read books about you.
I'll read 'cause I love you.
I'll read books about you.

I'll WRITE TO POLITICIANS.
I'll write to politicians.
I'll write 'cause I love you.
I'll write to politicians.

I'll ALWAYS BE GREEN.
I'll always be green.
I'll be green 'cause I love you.
I'll always be green.

I'll GIVE YOU RESPECT.
I'll give you respect.
I'll respect you 'cause I love you.
I'll give you respect.

Happy Earth Day to you.
Happy Earth Day to you.
You're MY HOME, and I love you.
Happy Earth Day to you.
MOTHER EARTH, how I love you,
HAPPY EARTH DAY TO YOU!

V. EARTH SCIENCE

WHAT IS YOUR BIRTHSTONE?

A birthstone is a glistening, gorgeous gemstone,
A cut and polished mineral of your very own.
Valued for its sparkling splendor and perfection,
Your gem is part of a twelve-month stone selection:

JANUARY is a GARNET that is deep red.
FEBRUARY is a purple AMETHYST instead.
MARCH is AQUAMARINE, a very pretty pale blue.
APRIL's the white DIAMOND you wear when you say, "I do."

MAY is a green EMERALD, like the green Irish isle.
A white PEARL on a necklace for JUNE makes a girl smile.
A JULY red RUBY sits in a dazzling ring.
A pale green PERIDOT in AUGUST for a king.

SEPTEMBER is a SAPPHIRE, a sparkling deep blue.
OCTOBER is a multicolored OPAL hue.
Yellow TOPAZ in NOVEMBER, like the leaves that fall.
DECEMBER's TOPAZ is blue and not yellow at all.

HOW DO YOU IDENTIFY A MINERAL?

How do you identify a mineral? What do you do?
Well, the OUTSIDE COLOR is one property or clue.
But weathering can make the outside color change some,
So you must use more clues, which is the rule of thumb.

Scratch testing for HARDNESS helps identify a mineral or a gem.
Choose a number on *Moh's Scale of Hardness* from one to ten.
The STREAK COLOR appears when the mineral's scratched on a special plate.
This color never changes, which makes this test helpful and first-rate.

Another clue is the way light reflects off a mineral to your eyes.
It gives the stone LUSTER, like shiny or dull. Using this clue is wise.
A fifth clue is CRYSTAL SHAPE or cleavage—how smoothly it breaks.
Breaking the mineral to find its crystal structure is all that it takes.

So, which mineral is it? It's *colorless and clear.*
Octahedral is its crystal shape, and it's *shiny* and *sheer.*
It's a *ten* on the hardness scale and *breaks the plate* before leaving a streak.
A spectacular *diamond*, the precious mineral that people seek.

WHO LET THE ROCKS OUT?

(This poem can be sung to the "Who Let the Dogs Out?" tune.)

Who let the rocks out … who, who, who, who?
Who let the rocks out … who, who, who, who?
Who let the rocks out … who, who, who, who?
Who let the rocks out?

Earth's pressure pushes magma out of the chamber.
The magma goes up into the vent.
More pressure pushes magma out of the crust.
It cools and forms rocks called igneous, igneous!

(Repeat verse 1.)

Pieces of sediment flow down the river.
Then they settle all the way down.
Years of pressure cement them together.
SEDIMENTARY rocks are now born, are now born!

(Repeat verse 1.)

Heat and pressure turn one rock to another.
Their physical form makes a change.
Igneous and sedimentary become METAMORPHIC.
This change takes many millions of years, millions of years!

(Repeat verse 1.)

WHAT ARE TECTONIC PLATES?

If you ask, "What are tectonic plates?" you're asking a very good question.
Many people have no idea, so they ask with apprehension.
Let's first explain what PLATES are to make it all clear.
They're HUGE CHUNKS OF THE EARTH'S CRUST AND UPPER MANTLE
called the *lithosphere.*

Earth's plates kept breaking apart at the end of dinosaur life,
Moving away from each other, causing the dinos great strife.
Thus, TECTONIC is the SLOW MOVEMENT OF EACH CRUSTAL PLATE,
Which is how the supercontinent *Pangaea* began to separate.

Plates can be made of land, or covered by ocean and have some of each.
Earth has eight big plates and many small—it's what geologists teach.
Plates can TRANSFORM past each other, crash, or CONVERGE,
And plates can move away from each other or DIVERGE.

Plate tectonics explains a lot of natural phenomena on Earth,
Like the huge mountain formations, starting from their birth.
These moving chunks of crust can also cause a lot of fear.
EARTHQUAKES and VOLCANOES create danger if you're near.

I'M A COMPOSITE CONE!
I'M A CINDER CONE!
I'M A SHIELD CONE!

ANYONE FOR A CONE?

Anyone for a cone? I mean, a large volcanic cone!
There are three types that you can find in a volcanic zone.

A SHIELD CONE is broad and low with shallowly sloping sides;
It has thin and runny lava that pours out slowly, and down it slides.

A CINDER CONE is taller than a shield and has sides that are steep;
It's made of hard, dry magma pieces that are many layers deep.

The third CONE is COMPOSITE, the largest of them all;
This type is a mountain, so it's always very tall.

PLEASE DON'T BLOW YOUR STACK

Please don't blow your stack, oh, magnificent volcanic cone!
I don't want to hear your loud noises or thunderous moans.
I know that your three cone types exist in different volcanic zones,
And two of them have dramatic eruptions that belch out hot stones.

My dear wide and low SHIELD CONE, you are so quiet and sweet.
Your thin lava streams out calmly—to watch you would be a treat.
You hill-sized CINDER CONES with steep sides give a more dramatic
  display
Of gas-filled lava chunks that shoot into the air and fall hard and gray.

Oh, COMPOSITE CONE, you are the last and largest type.
You're a majestic snowcapped mountain, but I just have one gripe:
Your thick and gas-filled lava that you belch out with all your might,
Causes violent explosions, so I think I'll stay out of sight.

WHAT'S SHAKIN'?

What's shakin'? What is shakin'?
I think the earth is quakin'!
Everything is jostling around,
And I hear a loud and frightening sound!

I live near the SAN ANDREAS FAULT LINE,
Which is a big clue and a sure sign,
That a big EARTHQUAKE is happening now.
I know exactly what, why, and how.

You see, a FAULT LINE is a crack in the earth's crust.
If you live near one, being prepared is a must.
If the PLATES CONVERGE or SLIP PAST EACH OTHER,
VIBRATIONS make things move or crash into one another.

The EPICENTER is the strongest point of the vibration or the earthquake.
The RICHTER SCALE numbers from one to ten show the strength of the shake.
AFTERSHOCKS are shock waves that happen after the main vibration.
They can occur over many days, so they create trepidation.

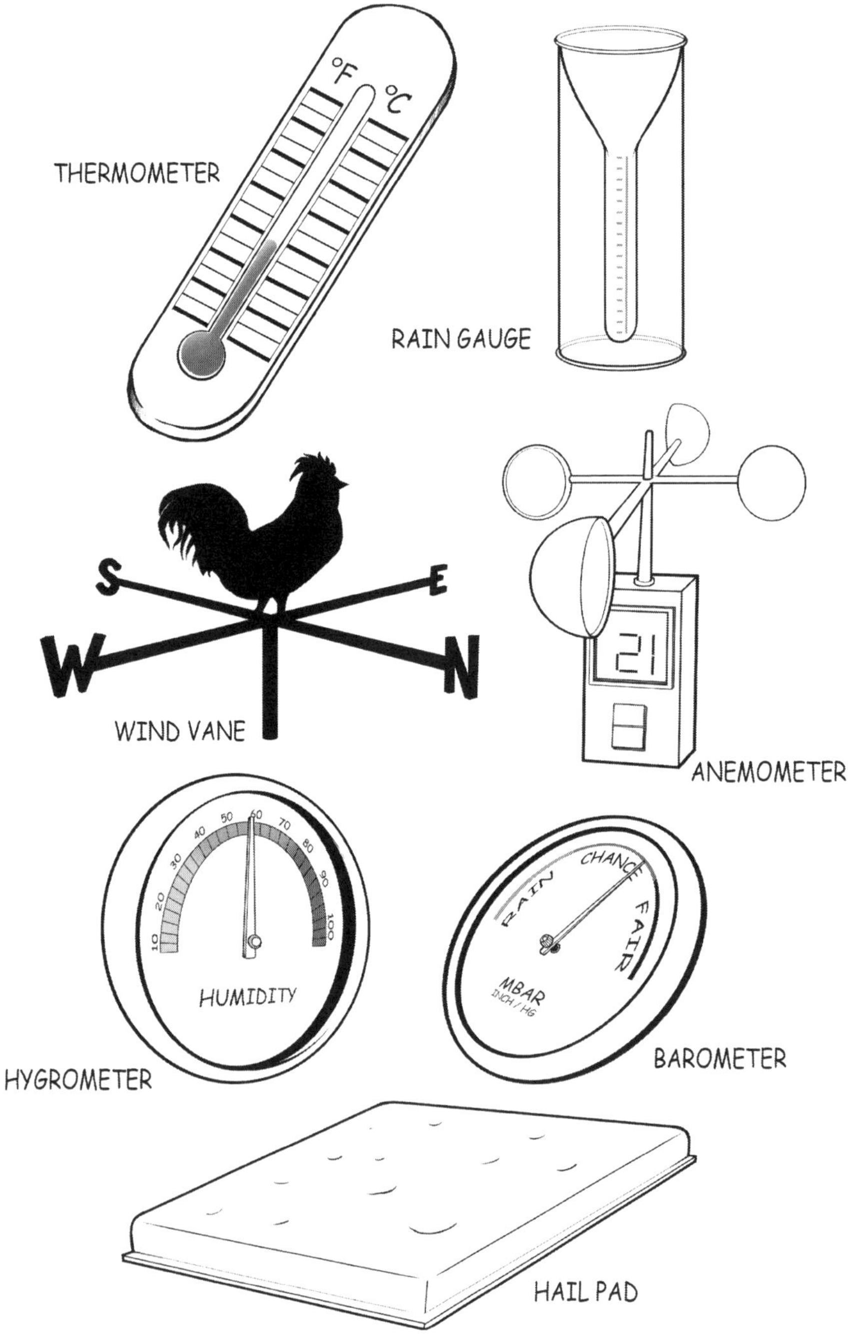
THERMOMETER
°F
°C
RAIN GAUGE
S
E
W
N
WIND VANE
21
ANEMOMETER
10 20 30 40 50 60 70 80 90 100
HUMIDITY
HYGROMETER
RAIN
CHANGE
FAIR
MBAR
INCH / HG
BAROMETER
HAIL PAD

WEATHER INSTRUMENTS

The weather affects everyone's life, and we have no control of it.
Meteorologists follow and predict it for our benefit.
Using various instruments, they measure every weather condition.
Then they quickly report their findings to the public, with great ambition.

THERMOMETERS are used to measure the temperature in degrees.
ANEMOMETERS measure the speed of a gusty wind or light breeze.
HYGROMETERS measure the amount of humidity that's in the air.
WIND VANES find the wind direction in stormy weather, or when it's fair.

Rain and snow are collected and measured in RAIN GAUGES.
BAROMETERS measure air pressure in millibar stages.
HAIL PADS are made of florists' foam and covered with aluminum foil.
The hail creates a measurable dimple before it will recoil.

A couple of other weather instruments are high-tech and more complex.
SATELLITES track weather systems over the large areas that it checks.
The DOPPLER RADAR is very helpful in detecting a large storm.
It determines the air circulation patterns when it's cold or warm.

THE WATER CYCLE GOES ROUND AND ROUND

(This poem can be sung to "The Wheels on the Bus" tune.)

The water cycle goes round and round,
Round and round, round and round.
The water cycle keeps going around:
LIQUID WATER ROUND TO WATER VAPOR.

LIQUID WATER EVAPORATES,
Evaporates, evaporates.
Heat speeds up the water molecules,
Then they jump into the air as vapor.

The WATER VAPOR CONDENSES,
Condenses, condenses.
Cooling slows down the vapor molecules.
They come together and form water droplets.

The water cycle never ends,
Never ends, never ends.
The water cycle goes round and round:
EVAPORATION ROUND TO CONDENSATION.

CLOUDS

Clouds are made of tiny droplets of water in the air,
Or itsy-bitsy ice crystals when clouds are way up there.
Hanging on to bits of dust and particles in the atmosphere,
Till they get too heavy and fall as precipitation down here.

Clouds form at different ALTITUDE LEVELS, like low, mid, and high.
How they reflect sunlight gives them different COLORS in the sky.
During the day, clouds can be white, or they can be light to dark gray,
Or even reddish orange during sunset at the end of the day.

Clouds also have a variety of SHAPE classifications:
They can be layered, heaped, curled, or even a combination.
The four cloud formations that are well-known to us
Are STRATUS, CUMULUS, CIRRUS, and CUMULONIMBUS.

All clouds are good indicators of the weather that will unfold,
Low gray STRATUS means LIGHT RAIN, and high curly CIRRUS means COLD.
Puffy CUMULUS always indicate a SUNNY AND FAIR DAY,
But watch out for big CUMULONIMBUS 'cause a STORM is on its way.

VI. PHYSICAL SCIENCE

ENERGY, ENERGY, ENERGY

Energy, energy, energy is always all around you.
It's even inside you when you're at rest and when you move too.
POTENTIAL and KINETIC ENERGY are the two main energy types.
The first is stored, and the second's in motion, like water running through pipes.
Within each group, there are six types that never quit.
That's why energy is everywhere, though you might not see it.

One ENERGY type is MECHANICAL, or what moving objects make.
Parts moving in a machine, like a mixer mixing batter for a cake.
Another kind is ELECTRICAL that's useful to us every day.
It's found in batteries and outlets that we use when we work and play.
RADIANT is next, giving energy from HEAT AND LIGHT,
Like the warm, bright sun in the day and a lightbulb at night.

SOUND ENERGY travels in waves when an object's vibrations are fast or slow,
Like vocal chords when you speak, and a musical instrument you strum or blow.
CHEMICAL ENERGY is released when there's a chemical reaction.
Strike a match and create a change from an interaction.
Last is NUCLEAR ENERGY, which is released in a nuclear reactor.
Splitting or fusing uranium atoms is a very important factor.

WHAT IS THE MATTER?

What is the matter? Is the matter a SOLID, a LIQUID, or a GAS?
It depends on the molecule arrangement and if they're slow or fast.

In a SOLID, molecules are packed tightly, and they can't move very much
'Cause every time the molecules move, they bump each other and touch.

Molecules are spread out in a LIQUID, so they're quick 'cause they're free.
They bump less than solid molecules 'cause they have more space, you see.

Molecules bump the least in a GAS, 'cause they're spread out with the
most space,
So gas molecules move more freely and quickly, as if they're in a race.

CHEMICAL AND PHYSICAL CHANGES

A CHEMICAL CHANGE is permanent, so it's irreversible too
'Cause its chemical composition changes to form a substance that's new,
Like wood to ashes from burning and iron to rust from oxidation.
The ashes and rust can't reverse back, which can be a frustration.

But a PHYSICAL CHANGE is only a change of one's state of matter,
Like water to vapor or ice and back to water that can splatter.
So, you see that after a physical change does what it does,
The object is then able to go back to what it once was.

WE'VE GOT CHEMISTRY

We've got chemistry—the study of a molecule's structure inside.
ELEMENTS have one atom type inside the nucleus or outside.
There can be more than one of this atom type, but there can never be
A different kind of atom inside an element, and that's the key.

The *periodic table of elements* lists one hundred eighteen.
The chemical formula of each and every element can be seen,
Like oxygen … O, hydrogen … H, sulfur … S, and carbon … C,
Or nickel … NI, zinc … ZN, gold … AU, and iron … FE.

Combine elements together, and a COMPOUND is what you get.
Numbers show the atom's amount; no number means one, don't forget.
One water molecule is H2O, and one ammonia equals NH3.
CO2 is carbon dioxide, and C12H22O11 is sucrose, you see.

SOUND AND LIGHT

We use sound and light during the day and night,
But do the two share the same traits? No, not quite.
They're both an important form of energy,
And they both travel very fast, you see.
But light can travel a lot faster than sound.
It travels faster than anything around.

Also, sound and light both travel as a wave.
The difference is *how* they travel and behave.
Vibrations make sound waves that need molecules to bump,
But light waves don't need molecules that move or jump.
So, sound occurs where there are molecules - no other place,
But light waves can travel everywhere - even in space.

They both travel in straight paths, and they both can bend.
However, light has only straight waves that it can send.
But water bends light waves as they move from the air,
And lenses bend light, like the glasses that we wear.
Both can pass through gas, liquid, and solid matter,
But light can't pass through opaque types of the latter.

THE HEAT IS ON

The heat is on, and it's traveling in different ways:
Through solids, liquids, gases, and also waves as rays.

CONDUCTION has molecule collision when a warmer and cooler solid touch.
The heat flows from the warmer to the cooler when you touch a hot pan or such.

When warmer areas rise to cooler in a liquid or a gas,
CONVECTION transfers heat in a circulation pattern through the mass.

When the transfer of heat doesn't need objects to touch, that's RADIATION.
The heat radiates through space as thermal rays, which is the explanation.

LIFE WITHOUT ELECTRICITY

Can you imagine life without electricity?
It is amazing to see how different life would be.
Electricity is intertwined with almost everything we do,
And it makes our life more comfortable, more fun, and easier too.

Just plug in your stereo, listen to a popular tune
On the radio or a CD to relax in the afternoon.
Watch your favorite sitcom on your flat-screen TV,
Or DVR a movie while you watch a video or DVD.

Connect a fun game to your TV and play it on the screen,
Or choose a game on your computer that is fast, fun, and clean.
And where would we be without the worldwide Internet,
To quickly google information that we need to get?

Turning lights on in our home or on our Christmas tree.
Microwaving a snack to eat after school at three.
And let's not forget about heat on a cold winter day,
Or ceiling fans to cool us when it's hot after we play.

This poem just shows a small sample of how life would be
Without the discovery of electricity.
So, we should appreciate that it's part of modern living.
An invention that will keep improving and keep on giving.

THE ATOM IS ALL CHARGED UP

The *atom is all charged up*, and I will tell you exactly why.
It's made of *particles with electric charges* that *some supply.*

In the atom's center, there might be *particles with no charge at all.*
These *neutron* particles are *neutral*, found inside the nucleus wall.

Besides neutrons in the nucleus, there might be PROTONS too.
These have a POSITIVE CHARGE, and moving slightly is what they do.

Most important are the ELECTRONS, which have a NEGATIVE CHARGE.
They quickly zip around the nucleus because their space is large.

If the atom has MORE PROTONS, then its overall charge is POSITIVE.
If the atom has MORE ELECTRONS then its total charge is NEGATIVE.

The *atom is neutral* or has *no charge* if the *protons and electrons are equal.*
Moving electrons produce electricity, but that's another poem or sequel.

I'm EDISON's PRIDE!
I SAVE ENERGY!

LIGHTBULB, LIGHTBULB

Lightbulb, lightbulb, light up for me.
Lightbulb, lightbulb, help me to see.

Lightbulb, lightbulb, watts are inside.
Lightbulb, lightbulb, Edison's pride.

Lightbulb, lightbulb, spiral's the key.
Lightbulb, lightbulb, save energy.

Lightbulb, lightbulb, you're so cute too.
Lightbulb, lightbulb, how I love you!

THE Cs OF ELECTRICITY

The form of energy known as electricity
Has a lot of cool words that start with the letter *C*,
Like CURRENT, the speedy negative electron flow,
And its path or CIRCUIT with a voltage strength high or low.

But without a CONDUCTOR, these charged electrons will surely stop.
They need metal or water to zip through, or the circuit will flop.
If the conductors are CONNECTED and all the circuit parts are good,
You'll have a CLOSED or COMPLETE circuit that will be working as it should.

Lightning is a
form of
ELECTRICITY!

WATCH THAT BOLT, BEN

Watch that bolt, Ben, 'cause LIGHTNING IS A FORM OF ELECTRICITY.
That's why you saw a spark and felt a zap when you touched the metal key.
Though it was dangerous to go out on that stormy day in 1752,
You were cautious, and your KITE EXPERIMENT proved your
hypothesis to be true.
You predicted that an object might receive lightning's electrical effect.
You connected wire from a key on a kite to a jar, where charges did collect.
The kite's wet, silk thread carried NEGATIVE CHARGES FROM THE
CLOUD TO THE KEY,
And taking cover, holding the thread's dry end, electricity you did see.
After this discovery, you created a LIGHTNING ROD, and you found
That atop a building, it carries charges through a wire down into the
ground.

WHAT DOES A GENERATOR GENERATE?

What does a generator generate, and exactly what does it do?
It generates a strong electric current that's quickly sent to you.
It begins with a twirling steam TURBINE, which makes a large central
SHAFT spin,
And the shaft turns a huge COIL OF WIRE through a MAGNETIC FIELD
therein.
This creates moving electrons or a current of electricity
That travels through long power lines to your house very rapidly.

MAY THE FORCE BE WITH YOU

May the *force* be with you. May the force be with *you*,
And with all your friends and your family too.
The force we get from the EARTH'S INVISIBLE MAGNETIC FIELD
Is the same force that attracting and repelling magnets yield.

Without this invisible field, there would be no magnetic force.
With no magnetic force, magnets wouldn't work at all, of course.
Our lives wouldn't be the same if we had no magnetic action
Because things we use each day need a magnetic field reaction.

We use compasses with magnetic needles to help us find our way,
And ride in cars, which run on generators with magnets, every day.
Computers and stereo speakers have strong magnets inside.
Microwaves and vending machines also have magnets that hide.

MAGNETIC MAGNETS

Magnetic magnets exist because of the EARTH'S MAGNETIC FIELD.
Without this invisible field, only pieces of iron would be revealed.
The Earth's magnetism creates the same field around magnets all the time,
Which makes the magnet attract to iron but not metals like a dime.
At EARTH'S TWO POLES, the MAGNETIC ENERGY is very STRONG,
And to all magnets, they pass this pole energy along.

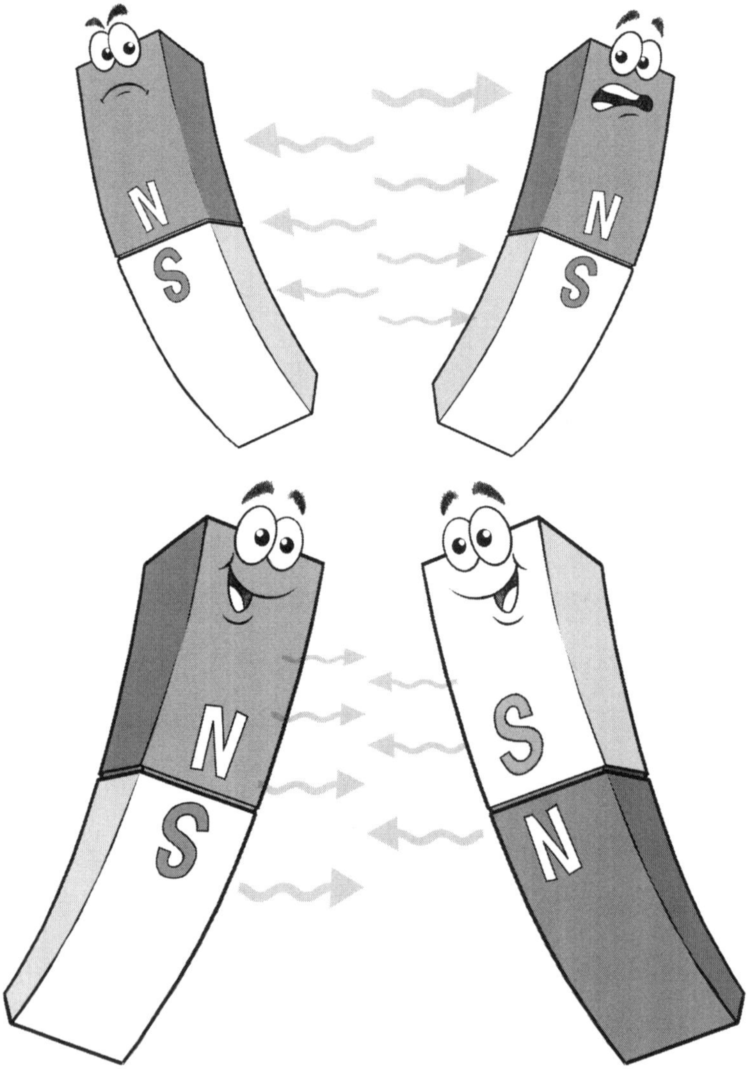
N
S
N
S
N
S
S
N

THE TWO MAGNETS HAVE A STRONG ATTRACTION

The two magnets have a strong attraction, and tell me, why shouldn't they?
A NORTH POLE AND A SOUTH POLE face each other and PULL with no delay.
People say, "Opposites attract," and with magnets, this is always true,
So if OPPOSITE/UNLIKE POLES ATTRACT, then what do same/like poles do?
LIKE POLES PUSH AWAY from each other, or in other words, they REPEL.
They do not like each other, so they're not attracted, and they don't gel.

I am a PULLEY!
I am an
INCLINED PLANE!
I am a SCREW!
I am a LEVER!
I am a WEDGE!
I am a
WHEEL AND AXLE!

SIX SIMPLE MACHINES

There are six simple machines that we use every day.
They help us work in an easier and faster way.

An INCLINED PLANE is a ramp or surface slanted big or slight.
It's mainly used to *roll an object up to a greater height.*

The next simple machine is the very common spiral SCREW.
A screw *holds objects together* in most things we have or do.

A WEDGE is the third machine that's used *to cut or separate,*
Like metal axes, knives, and needles that are so sharp and straight.

Fourth is a LEVER that *moves objects* at its fulcrum in each mode.
The three types or classes are based on the location of the fulcrum
and the load.

The fifth is a WHEEL AND AXLE, like a tire or Ferris wheel.
A wheel moves around a rod, *helping to move things.* It's ideal.

Last is a PULLEY used to *lift or move objects* with control.
One of the very most common pulleys is a tall flagpole.

Without these simple machines, life would be harder, no doubt,
Because we'd use more *effort force* and that would make us pout.

COMPOUND MACHINES

A compound machine is TWO OR MORE SIMPLE MACHINES.
When simple machines come together, a useful tool is seen.
Compound machines are complex, 'cause they have more than one part,
And they *help us do more difficult work*, straight from the start.
The compound machines are the type of machines that *we use the most.*
They are used every day by the young and the old from coast to coast.

We use these compound machines to do work and also when we play.
Here is a sample of compound machines that we use every day:
A lever and wedge combine to make SCISSORS or a GARDEN HOE,
And two big levers and a wheel and axle make a WHEELBARROW.
A BICYCLE is one of the most complex machines that we use.
It has wheels and axles, levers, a pulley, and a whole lot of screws.

THE TWO FORCES I LOVE

FRICTION and GRAVITY are the two forces I love.
One's on the ground, and the other's above.
With no friction, I'd slide all over the place,
And with no gravity, I'd float right up into space.
So thank you, two forces that I love so,
For your help and protection, as through life I go.

This is YUMMY!
This is NUTRITIOUS!
milk
milk

VII. HEALTH SCIENCE

NUTRITION AMBITION[1]

Make it your daily ambition
To eat foods with nutrition.
Having a BALANCE is the key,
And a lot of VARIETY.

The WHOLE GRAIN pastas, breads, and rice
Smell and taste yummy and entice.
You also need to have PROTEIN,
Like eggs, nuts, beans, and meat that's lean.
Have some orange and green VEGGIES on your plate;
Their VITAMINS AND MINERALS are great.
And LOW-FAT DAIRY is good for you,
So have milk, yogurt, and some cheese too.
FRESH FRUIT makes a really great snack,
And helps you stay on the right track.

Use PORTION CONTROL every day,
And a little dessert is okay.
A little of this and a little of that,
Even SOME SUGAR and SOME FAT.

---

1 Be careful! You might have a food allergy to foods like nuts, eggs, or even dairy. So, check with your mom if you're not sure of certain foods, even though they're healthy and pure.

THIS IS FUN!

EXERCISE FOR FITNESS

Exercise for fitness every day.
Run, bike, or jump rope for outside play.
It's fun to ice-skate and roller-skate,
Or Roller-blade at the rink till eight.

Ride on a skateboard or on a scooter,
Instead of sitting at your computer.
Dance to some music, and it's no surprise,
That it's fun and also good exercise.

Enjoy games of baseball or softball,
Basketball, soccer, and football.
A singles or doubles tennis game.
Doing activities is your aim.

It's great exercise to go for a swim
'Cause it will help you be healthy and trim.
Sometimes it's fun just to take a walk.
Go with a friend and have a nice talk.

These aerobics are all good for your heart
'Cause they get the blood flowing from the start.
They also give you lots of energy,
And that is a good, healthy guarantee.

DOES IT MAKE SENSE TO HAVE FIVE SENSES?

Does it make sense to have five senses?
Do we really need to have all five?
Yes, we need them for our protection
To keep us healthy and alive.

We use our TONGUE for TASTING foods and drinks we love,
And also to avoid eating spoiled food.
SMELLING with our NOSE to smell old food that is rotten,
And to alert us of fire's smoke we need to elude.

With our EARS, we HEAR music and nature's lovely sounds.
And they warn us that danger's near when we hear an alarm.
We SEE the rainbow colors with our EYES and the smiles of friends.
They also help us see hazards, keeping us from peril and harm.

We use our sensitive SKIN as our fifth sense, which is TOUCH.
We feel the heat of a stove and the wetness of morning dew.
So, yes, we need all our five senses as we go through each day.
Thank goodness for our senses, which help us survive and have fun too.